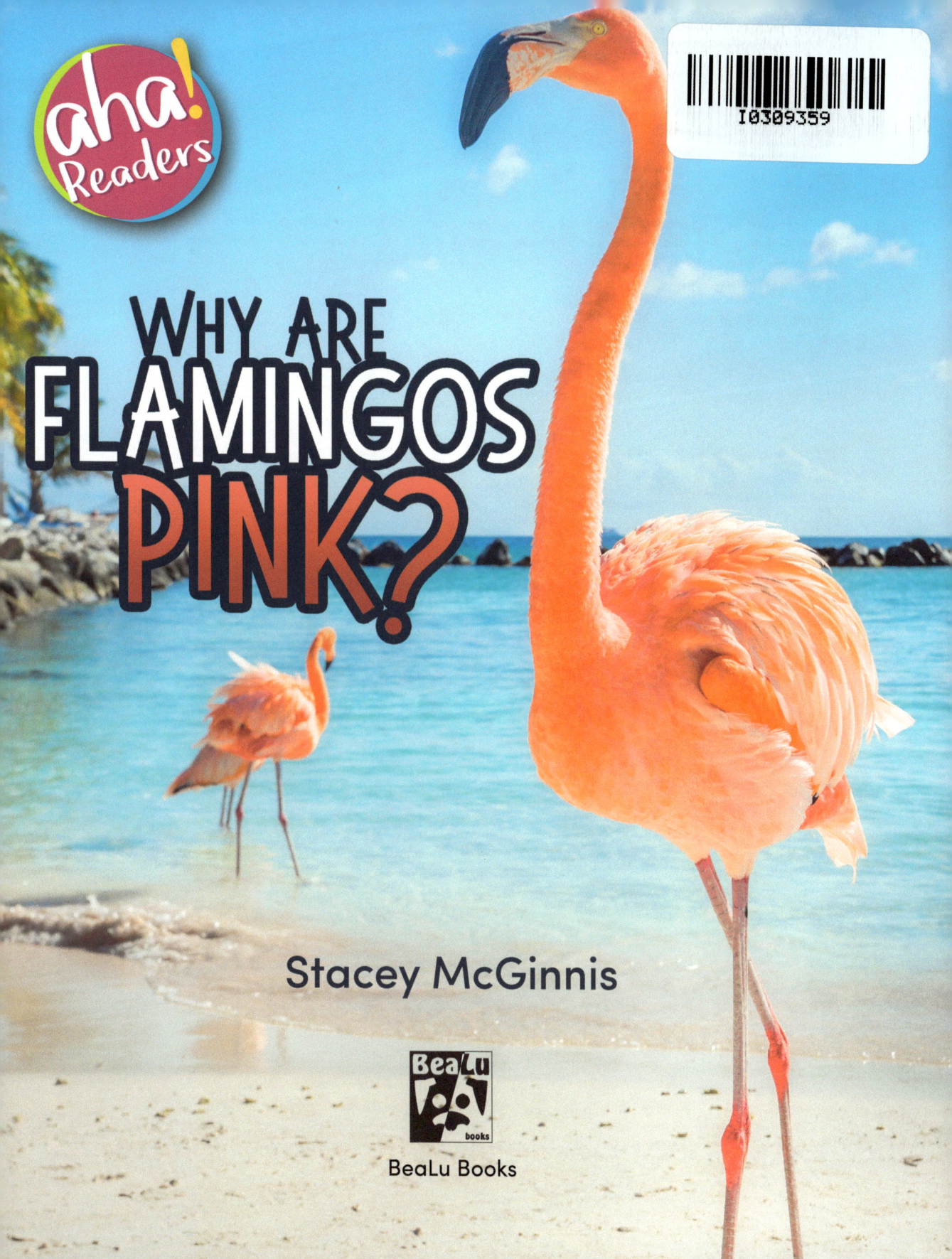

Copyright © 2020 by Stacey McGinnis

All rights reserved. No part of this publication may be reproduced in any form or by any electronic or mechanical means, including information storage and retrieval systems, without the express written permission from the publisher, except in the case of brief quotations embodied in critical articles or reviews. For information regarding permission, contact BeaLu Books.

ISBN Hardcover: 978-1-7341065-5-8
ISBN Paperback: 978-0-9990924-7-7

Library of Congress Control Number: 2019952467
Publisher's Cataloging-in-Publication Data is on file with the publisher.

Edited by: Luana K. Mitten
Book cover and interior design by Tara Raymo • creativelytara.com

Printed in the United States of America
October 2019

BeaLu Books
Tampa, Florida

www.BeaLuBooks.com

PHOTO CREDITS: Cover: © FER737NG; Page 1: © MasterPhoto; Page 3: © Gints Ivuskans, © pzAxe, © Firefighter Montreal, © Jevanto Productions; Page 4: © PSmith USA, © Ye.Maltsev, © Manfred Ruckszio, © Tree Vonguitarat; Page 5: © PLANET EARTH; Page 6: © Salparadis, © Eric Isselee; Page 7: © Vladimir Sazonov; Page 8: © David ODell, © Eric Isselee; Page 9: © La Nav de Fotografia; Page 10: © Tish1; Page 11: © tab62, © David A. Litman; Page 12: © Baishev; Page 13: © bcampbell65; Page 14: © wildestanimal, © Junlian Gunther; Page 15: © Fine Art Photos; Page 16: © Stas_Shvalor, © Ondrej Prosicky; Page 17: © Prime Photo, © Brandon Alms, © Nicola Auckland

Color is all around us. Sometimes a color tells us when to stop or go. Sometimes color is a warning of danger. Color also has special meanings or purposes in nature.

WHY ARE FLAMINGOS PINK?

algae

brine shrimp

larvae

Flamingo is a Spanish or Portuguese word meaning flame-colored.

The saying, "You are what you eat" is true for flamingos. They get their pink color from the algae, larvae, and brine shrimp they eat. The bright pink color comes from beta-carotene, which gives things a red, orange, or yellow coloring. Foods we eat such as carrots, sweet potatoes, spinach, kale, and apricots contain beta-carotene too.

WHY DO GIRAFFES HAVE BROWN PATCHES?

The patches on the giraffe are made up of many blood vessels that act as panels to release heat and help cool the giraffe.

Taller than a skyscraper is a giraffe. No, not really! But giraffes are the tallest land animal. Giraffe's patches protect them by providing camouflage from predators and by helping them stay cool in the hot temperatures of the African savanna.

WHY IS THE POISON DART FROG RED?

Poison dart frogs aren't just red. They can be yellow, gold, copper, green, blue, or black. But female poison dart frogs prefer the red male frogs.

The bright red color of the poison dart frog means, "Stay back!" When predators see red, they know not to eat this poisonous frog. Even though they have "poison" in their name, these frogs can't make their own poison. Instead, they get it from the toxins of ants and mites that they eat.

WHY IS A BUMBLEBEE YELLOW AND BLACK?

There are over 20,000 varieties of bumblebees and honeybees in the world.

Yuck! The yellow and black color tells predators that bumblebees are not tasty. Be careful if you decide to provoke a bumblebee. They can sting multiple times without dying. Bumblebees live in nests often found in burrows or holes in the ground. So be careful where you walk!

WHY ARE ALLIGATORS GREEN?

The largest America alligator officially weighed and measured was caught in the Alabama River in 2014. It measured 15 feet, 9 inches long (4.8 meters) and weighed about 1,012 pounds (459 kilograms). Now that's a big gator!

Alligators skin color comes from where they live. The algae in water stick to their skin and make them look green. Give an alligator a bath and scrub a dub, they're grey. When leaves fall in the water and start to decay, they create tannic acid. Alligators who live in water with lots of decaying leaves may have more brown, gray, or nearly black skin.

WHY ARE POLAR BEARS WHITE?

Polar bears like to keep clean by swimming in the water or rolling in the snow of their home in the Arctic.

Polar bears look white but guess what? They're not! Their skin is black, and their fur is hollow tubes that are transparent or see-through. They look white because the hollow tube-like fur scatters light. The white color allows polar bears to blend into their environment. This lets them sneak up on their prey.

WHY ARE ELEPHANTS GRAY?

One elephant can eat 300 pounds (136 kilograms) of food a day. That would be like drinking 33 gallons (125 liters) of milk in one day.

Elephants grey color is a mix of dark and light colors. This color combination allows an elephant to get rid of heat in the summer and absorb heat in the winter. Dark colors absorb heat, and light colors reflect heat. Sometimes elephants look brown or reddish when wallowing in mud.

WHY ARE PACIFIC TREE FROGS DIFFERENT COLORS AND PATTERNS?

Pacific tree frogs are only 1 to 2 inches long (2.54 to 5.08 centimeters).

Most Pacific tree frogs are either green or brown to start, but they can change color and patterns, too. These frogs may change color or pattern to hide from danger. They also change color to help maintain their body temperature. When temperatures are warmer, the Pacific tree frog will often turn a shade of yellow. And when cooler weather comes, they can turn back to green and brown.

WHY IS A CHAMELEON GREEN? BLUE? ORANGE?

The 160 species of chameleons found in Africa, Asia, Europe, and North America are all sizes. The longest chameleon is the Madagascan and can be up to 23 inches long (58 centimeters). That's the length of two sheets of paper.

Wait, chameleons aren't just green, they can be almost any color. We used to think that chameleons changed color for camouflage. Scientists now believe the color change is less about safety and more about showing their moods. Darker colors can show that they are angry. Lighter colors help to attract a mate. Just like elephants and Pacific tree frogs, chameleon's change to darker colors to absorb heat.

WHY IS THIS OCTOPUS BLUE?

When an octopus feels threatened, it will shoot a dark inky cloud into the water so it can hide and then getaway.

Yes, it's true! An octopus can change its color to match its surroundings. They will take on the colors blue, brown, pink, grey or even green to hide from predators or sneak up on prey. Some octopi even use color change to communicate with each other.

WHY ARE SOME SHARKS BLUE AND WHITE?

Some sharks can go through 35,000 teeth in a lifetime. That's good news for a dentist!

Sharks come in many shapes, sizes, and patterns. One thing they all have in common is a darker colored body with a lighter colored stomach. When looking at a shark from above, the darker color helps them blend into the water. When looking up at a shark from below, the lighter coloring helps them blend into the surface of the water. This type of coloring is a form of camouflage called countershading.

WHY DOES A SEA TURTLE HAVE A **DARK** SHELL?

Unlike turtles we see on the land, sea turtles cannot retract, or pull back, their heads into their shell.

Besides being neighbors in the ocean, sea turtles and sharks have something else in common. They both use countershading for camouflage. While sea turtles come in many colors and sizes, their shells are often darker in color, and their stomachs are light in color. This countershading protects them from predators from above and below. Some sea turtles, such as the hawksbills, have patterns on their shells.

Color is important in many ways, not only for humans but also for our animal friends! So next time you see an animal ask yourself, "Why is it that color?"

ABOUT THE AUTHOR:

Stacey McGinnis is a teacher at Trinity Oaks Elementary School in New Port Richey, Florida. She has been an elementary school teacher for 20 years. Writing is a passion for Ms. McGinnis. When she is not teaching or writing, she can be found trying new crafts and playing with her houseful of fur babies.

Learn more!

Besides being different colors, there are lots of other fun facts about animals. Can you find the "why" for these facts?

Did you know the heart of a shrimp is in its head!

Did you know that elephants can't jump!

Did you know that an ostrich's eye is bigger than its brain!

Did you know cheetahs have to drink only once every 3-4 days!

Did you know koalas sleep for up to 18 hours a day!

Did you know emperor penguins can stay underwater for up to 22 minutes!

www.ingramcontent.com/pod-product-compliance
Lightning Source LLC
Chambersburg PA
CBHW041443010526
44118CB00003B/161